PENGUIN POCKET GUIDES

# NEW ZEALAND'S NATIVE FLOWERS *of the* Bush

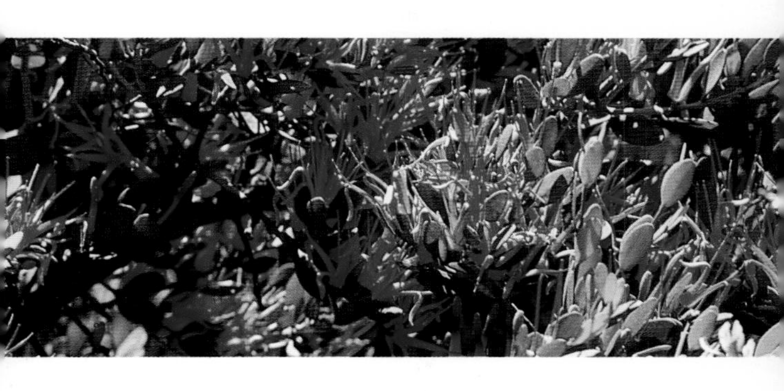

PENGUIN POCKET GUIDES

# NEW ZEALAND'S NATIVE FLOWERS *of the* Bush

Text by
Colin Ogle

PENGUIN BOOKS
Penguin Books (NZ) Ltd, cnr Rosedale and Airborne Roads,
Albany, Auckland, New Zealand
Penguin Books Ltd, 27 Wrights Lane, London W8 5TZ, England
Penguin USA, 375 Hudson Street, New York,
NY 10014, United States
Penguin Books Australia Ltd, 487 Maroondah Highway,
Ringwood, Australia 3134
Penguin Books Canada Ltd, 10 Alcorn Avenue, Toronto,
Ontario, Canada M4V 3B2

Penguin Books Ltd, Registered Offices: Harmondsworth,
Middlesex, England

First published by Penguin Books (NZ) Ltd 1997
1 3 5 7 9 10 8 6 4 2


Typeset by TTS Jazz, Auckland
Printed by Condor Production, Hong Kong

ISBN 0-140-26713-1

# CONTENTS

ACKNOWLEDGEMENTS

Photographs in this book are the work of the following. The publisher acknowledges their expertise and assistance: **John Barkla:** pp 1, 2-3, 20; **Brian Chudleigh:** pp 15, 18, 42, 54, 57; **Department of Conservation:** pp 5, 16, 19, 22, 23, 24, 25, 30, 31, 35, 43, 44, 55, 58; **Don Hadden:** pp 12-13, 21, 27, 28, 32, 37, 38, 49, 52, 56, 59; **Jan and Eroll Kelly:** pp 8, 36, 48; **Key-Light Image Library:** pp 17, 50; **Rod Morris:** pp 26, 29, 33, 34, 39, 40, 41, 45, 46, 47, 51, 53; **Craig Potton:** p. 14.

# USING THIS BOOK

This **Penguin Pocket Guide to New Zealand's Native Flowers of the Bush** is an illustrated guide to a selection of endemic and native flowering plant species found in this country.

The order in which subjects appear has been determined alphabetically by common name rather than by the more usual scientific classification of plant groups, the intention being to provide a work that can be more readily used by the general reader. A note on the plants' scientific classification appears on page 62.

The status of the species is based on the following definitions:

**Endemic** — occurring in the wild *only* in New Zealand.

**Indigenous** (Native) — occurring in the wild in New Zealand and in the wild outside this region.

# INTRODUCTION

The total number of flowering plant species native to New Zealand — species thought to have been here before human presence — is just over 2200. Of these, about 600 grow in native forest habitats. Although the number of native species is not high for a land mass of New Zealand's size and variety of land forms and climate, all but a handful of the species are endemic to New Zealand, meaning that they do not grow naturally in any other country.

Such a proportion of endemic species reflects New Zealand's isolation in time and space from sources of other plants. In prehistoric times, New Zealand was part of a single great land mass, the southern supercontinent of Gondwana. New Zealand separated from the rest of Gondwana about 80 million years ago, this drift away from the larger land mass occurring after its colonisation by amphibians, reptiles and birds, but before the southwards dispersal of mammals to the region. So, free from browsing mammals, New Zealand's plant life evolved a special and unique character.

Primeval trees and smaller plants remain to this day, and have their nearest relatives in other Gondwanan fragments, particularly Australia, South America and New Caledonia. Examples are the southern beeches (*Nothofagus*); the podocarps (ancient conifers that include rimu, totara, kahikatea and matai), araucarian conifers such as kauri, and the protea family which contains *Knightia* (rewarewa).

Blending with the southern species in New Zealand's native vegetation is a suite of plants whose ancestors are thought to have arrived more recently from sub-tropical lands, including Australia, since the Gondwana break-up. There are various theories on how they got here. The seeds of some may have 'island-hopped' by floating, as for example, mangrove, and kowhai; or by being windborne (many native orchids, daisies and ferns, for example); or carried by birds (poroporo, and the ancestors of puriri, hinau and kaikomako are possibilities).

New Zealand's bush is not all the same. What tree species predominate depends on factors such as climate, soil, latitude and the history of disturbance at the site. As a broad guide, New Zealand's bush can be classed as either coniferous forest (often separated into podocarp, cedar and kauri forests), or beech forest (dominated by one or more of the five types of native beeches), or broad-leaved forest. The last may be dominated by one tree species, such as pohutukawa (*Metrosideros excelsa*) or tawa (*Beilschmiedia tawa*), but is often more diverse and classed, for example, as pohutukawa-puriri, or rata-tawa-kamahi forest.

As well as having endemic species, the bush is distinctly 'New Zealand' for a combination of features. It is mostly rainforest, meaning that it has a lush appearance, even in cool parts of the country. This is because there is year-round rain which supports abundant ferns, vines and epiphytes, and the tree trunks and branches are frequently clothed in mosses, liverworts and leafy lichens.

The trees are evergreen, and several species have trunk buttresses or aerial roots. Tree ferns are widespread and often common. Many of these features give the bush a more tropical look than would be expected from the southerly latitude of New Zealand.

Although there are exceptions, as evidenced by this book, the trees generally lack large or colourful flowers.

About one third of New Zealand's bush was burnt or otherwise cleared between the arrival of Polynesians settlers, 800 to 10,000 years ago, and Europeans just two centuries ago. Another third has been cleared since. Most of this loss has been of lowland forest.

Some of the plants illustrated in this book were once much more widespread and common than they are now. Although loss of forest habitat is often a factor, browsing by introduced mammals (deer, goats, pigs and the Australian brushtailed possum being the most widespread and common) is nearly always the main cause of decline. Examples include native mistletoes (*Peraxilla colensoi*), kaka beak (*Clianthus puniceus*), and *Dactylanthus*.

Offshore islands seem to offer better survival prospects for certain native plants but nearly all bush plants can be maintained in their mainland bush habitats. What are needed are the human and financial resources to undertake and maintain control of wild animals and weeds, to fence out domestic animals and to teach people about protection and wise use of the remaining bush.

*Colin Ogle*

# NATIVE FLOWERS

*of the* Bush

Don Hadden

MOUNTAIN BEECH

## MOUNTAIN BEECH *(Nothofagus solandri* var. *cliffortioides)*

One of the five kinds of evergreen native beeches, mountain beech is sometimes recognised as a distinct species (*Nothofagus cliffortioides*) rather than a variety of *N. solandri*. In common with other beeches, the flowers of mountain beech are tiny, and either male or female, but are produced on the same trees. Only the male flowers are colourful and in good flowering years they can be so abundant that whole trees appear rusty-red. Male flowers are about 2 mm across, with 8-14 red anthers that disperse their pollen by wind. Female flowers are smaller and green, and are reported as ripening two or more weeks later than male flowers on the same twigs. The fruits are 6-7 mm long hard nuts. Mountain beech is often the dominant canopy tree in montane forest or a shrub in sub-alpine zones. It tolerates wet sites and low-fertility soils, where it typically reaches 15 metres tall, but can grow to 25 metres in favourable conditions.

ENDEMIC SPECIES

Key-Light

NATIVE CALCEOLARIA

## NATIVE CALCEOLARIA *(Jovellana sinclairii)*

Native calceolaria is one of the loveliest herbaceous plants to be found in the bush. In fact, New Zealand forests have a rather small number of herbs, especially ones with such eye-catching flowers as those of native calceolaria. Each flower is about 10 mm across, its white petals joined to form a goblet, spotted with purple inside and containing two short stamens and a single short style. Carried on fine flower stalks at the tip of erect leafy stems of up to 300 mm in length, the white flowers stand out against the dark foliage and general gloom of the forest floor, their visual impact multiplied because they grow on branched panicles of 10-20 flowers. Small dry capsules follow the flowers. The species grows between East Cape and the inland Hawke's Bay and Waiouru districts, usually on damp fertile river flats under forest. *Jovellana* is a small genus in New Zealand and Chile, closely related to *Calceolaria*, which is also South American. The only other native species is *Jovellana repens*, a smaller plant than *J. sinclairii*, and which grows mostly in semi-shade among rocks close to water.

ENDEMIC SPECIES

DOC/R. Morris

NATIVE CLEMATIS, PUAWHANANGA

## NATIVE CLEMATIS *(Clematis paniculata)*

The large (80-100 mm diameter), pure white male flowers of puawhananga make this the best-known native species of *Clematis*. As in other *Clematis* species, the eye-catching parts of the flower are not petals but sepals. Puawhananga flowers have 5-8 sepals, white on male plants and slightly greenish on females. Female flowers are about half the size of males. Male flowers have many stamens with mauve anthers, while females have many separate ovaries and styles, with an outer ring of abortive stamens. When a puawhananga vine reaches the tree tops, its bunches of flowers stand out against the green canopy, in early spring in the lowlands or as late as December in montane forests. Female flowers develop spherical heads of seeds with fluffy styles to assist dispersal by wind. Native *Clematis* species grow in open places, shrublands or forest edges, but puawhananga is also common in the heart of mature forest and grows throughout New Zealand. Adult leaves of puawhananga are thick and dark green, and divided into three leaflets without edge teeth or lobes; seedlings have remarkably variable leaf shapes.

ENDEMIC SPECIES

John Barkla

TREE DAISY

## TREE DAISY *(Olearia arborescens)*

*Olearia arborescens* is a round-headed tree, to about 5 metres, growing mostly in montane zones, on forest edges, river banks and shrublands. In summer, its panicles of white daisy heads are sometimes sufficient to hide most of the dull green leaves, and the tree becomes visible from afar. The musky perfume of the flowers can be striking on a warm, calm day. Like many daisies, each flower head (capitulum) comprises many florets (small flowers) of two main types, ray florets and tube florets. In *O. arborescens*, 7-10 ray florets are arranged in a ring and are conspicuous through having elongated strap-like white petals. The centre of each capitulum has 10-12 closely-packed tube florets with five minute tooth-like tips to their petals. There are about 35 species of native trees and shrubs in the daisy genus *Olearia*, ranging from coastal to sub-alpine habitats. A number of species hybridise in the wild, and others have been produced in gardens.

ENDEMIC SPECIES

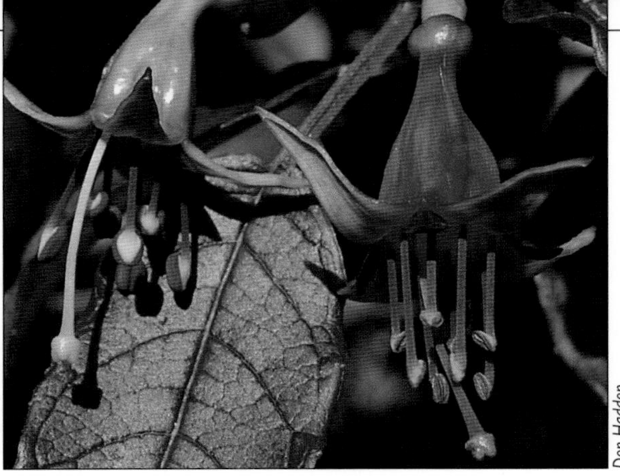

Don Hadden

TREE FUCHSIA, KOTUKUTUKU

## TREE FUCHSIA *(Fuchsia excorticata)*

With an unmistakable resemblance to the flowers of their garden fuchsia relatives, kotukutuku flowers are up to 30 mm long, with four sepals and petals, and eight stamens. The outer ends of the sepals are separate and petal-like but are joined into a tube in the basal half. The sepals, including the tube, are green when the flower opens, turning red later. The petals are barely a third the length of the sepal lobes, and are narrow and purple. Trees are either hermaphrodite, in which the stamens have long filaments and produce ample blue pollen, or female, in which the stamens have short filaments and no pollen. Both kinds of flowers produce black berries, about 10 mm long. The tree fuchsia is a wide-branching tree to 12 metres tall, with very distinctive peeling, papery, pink-fawn bark. In cool climates the trees are deciduous in winter. Although its abundance has been greatly reduced by the browsing of the introduced brush-tailed possums, the tree fuchsia is still a widespread lowland and montane tree on forest edges, river banks, erosion scars and deep in forested gullies.

ENDEMIC SPECIES

DOC

NEW ZEALAND GLOXINIA, TAUREPO

# NEW ZEALAND GLOXINIA

*(Rhabdothamnus solandri)*

New Zealand glozinia grows to about 2 metres tall as a loosely tangled shrub. It has roundish, toothed, bristly-hairy leaves in opposite pairs. The very distinctive orange-red, or sometimes yellow or red flowers can be found most of the year. Except for their spreading tips, the five petals are joined in a tube about 25 mm long. Two petal lobes are more closely joined than the other three, making a slightly two-lobed mouth to the flower. Four stamens are arranged with the anthers lying under the upper lip. There they can deposit pollen on the heads of birds that visit for nectar. The stigma ripens later and receives pollen brought by birds from other flowers. The fruit is a dry capsule with many seeds. New Zealand glozinia is found through much of the North Island, in well-lit places in coastal and lowland forest. It is a typical plant of stream banks, stony areas, cliffs and steep hillsides under the forest canopy, or semi-shaded places in the open. It can be abundant in limestone and mudstone country.

ENDEMIC GENUS

DOC

HINAU

## HINAU *(Elaeocarpus dentatus)*

A hinau tree in full bloom in late spring may not be obvious from afar, but close up it has an appealing simple beauty. White cup-shaped flowers of hinau droop in sprays of about 10 flowers, resembling flower spikes of lily-of-the-valley. Individual flowers are 8-15 mm across and have petals with 3-5 notches or lobes, like much enlarged flowers of wineberry (makomako); wineberry and hinau are in the same family. Flowers are followed by one-seeded purplish drupes, up to 18 mm long. Hinau is common in lowland and montane forests through the North Island and is scattered in lowland forests of the South Island to the latitude of Mt Cook in the west and south Otago in the east. Hinau grow to 15 metres or more and their trunks are easily recognised by their vertically grooved grey bark. The adult leaves are 10-12 mm long, widest near the tips and with serrated margins curved downwards. Hinau and pokaka (*E. hookerianus*) often grow together, although the latter also grows in cooler climates than hinau; hybrids are known between hinau and pokaka.

ENDEMIC SPECIES

DOC

FOREST IRIS

## FOREST IRIS *(Libertia pulchella)*

The small forest iris, *Libertia pulchella*, is a tufted grassy plant made up of several fans of stiff narrow leaves about 60 mm long. Its flowers are 10-15 mm across and grow on fine erect branches about 150 mm tall, each with up to 12 flowers. *L. pulchella* differs from other *Libertia* species in that the six flower tepals are of roughly equal size, rather than having the outer three much smaller than the inner three. Despite their small size, the pure white starry flowers are easily seen against the dark greens and browns of the forest floor. The fruit is a small brown capsule that splits into three when ripe to release its pale yellow seeds. *L. pulchella* grows in cool wet forests of the mountains. It is widespread in the North Island, but mainly in the west of the South Island and reaching sea level in Fiordland. Its habitat is a sharp contrast with the dry lowland habitats of its larger relatives (see *Libertia grandiflora*, p.38).

INDIGENOUS SPECIES

Rod Morris

KAKA BEAK, KOWHAI NGUTU-KAKA

# KAKA BEAK *(Clianthus puniceus)*

Although a very popular garden plant, kaka beak is one of the rarest native shrubs in the wild state. Lake Waikaremoana is often stated to be the last place with kaka beak in the wild, but it has also been found recently north of Gisborne and in inland Hawke's Bay, on steep land where browsing animals cannot reach it. The flowers hang in racemes, with the arrangement of floral parts being typical of the pea family. The 'top' petal (standard) is large and upright, the side petals (wings) are small, and the two 'bottom' petals are joined as a prominent beak-like keel enclosing the stamens and style. The flowers produce flattened black pods that split to release many small seeds. There is some variation between wild plants in flower size and petal colour, with the largest flowers being about 100 mm long. Pink and white-flowered forms are in cultivation. Kaka beak shrubs grow to about 2 metres tall, though they can be taller among other shrubs. Branches that bend to the ground can take root, resulting in colonies of genetically identical shrubs.

ENDEMIC SPECIES

Don Hadden

KAMAHI

## KAMAHI *(Weinmannia racemosa)*

From a distance, flowering kamahi forest looks to be dusted in snow. Kamahi flowers are small, but packed into 80-120 mm spikes (racemes) over the crown of the tree. Each flower has four or five white or pink petals, 2-3 mm long. The stamens are longer than the petals, and twice the number, making a raceme rather like a bottlebrush. Nectar-feeding birds pollinate the flowers and, today, introduced honey bees are also major users of kamahi nectar. Kamahi fruits are small dry capsules. The traditional view of *Weinmannia* in New Zealand is of two species, *W. silvicola* (towai) and *W. racemosa* (kamahi) which grow to the north and south of the western Bay of Plenty respectively, but some botanists regard these as one species. In some places, the name tawhero is used for either towai or kamahi. Kamahi is, perhaps, New Zealand's most abundant native tree. Growing to 25 metres, it can be the dominant tree, especially on low-fertility soils in high rainfall areas. In recent years, the death of much kamahi forest has been blamed on the browsing of introduced brush-tailed possums.

ENDEMIC SPECIES

Don Hadden

KANUKA

## KANUKA *(Kunzea ericoides)*

The beauty of massed flowers of kanuka in early summer is just one of this tree's many values which are under-rated when kanuka is referred to disparagingly as 'scrub'. A kanuka flower is about 5-7 mm across with five white or pale cream petals and five smaller, greenish-white sepals, visible from above in the gaps between the petals. These and 20-40 stamens line the edges of a shallow disc which holds nectar. Like those of its relative, *Eucalyptus*, the stamens of kanuka are folded inwards in the unopened bud. As the flower opens, the stamens straighten and elongate, although some inner stamens remain short and curved. The capsules contain many fine seeds and often fall before the next flowering. Kanuka can grow into a round-headed 20 metre tree with a trunk up to one metre across. It often grows as almost pure stands after fire or other disturbance. The small leaves contain oil glands. Manuka (*Leptospermum scoparium*) resembles kanuka, but has larger flowers and larger and more woody, persistent capsules; it is also a smaller and usually shorter-lived shrub.

ENDEMIC SPECIES

Rod Morris

KARAPAPA

## KARAPAPA *(Alseuosmia macrophylla)*

In spring, a sweet scent floating through the forest gloom often leads to flowers of karapapa. Close up, the scented flowers have five crimson petals joined as a trumpet 30-45 mm long, with five stamens inside and frilled petal edges. There are reports of nectar- feeding birds visiting the flowers, and moths are probably pollinators as well. The name *Alseuosmia,* meaning 'perfume of the glade', was created in 1839 by botanist Allan Cunningham. The number of *Alseuosmia* species is still doubtful because of the bewildering range of forms in some places. These may be hybrids or undescribed species, or both. Karapapa has the largest flowers and is a more or less upright bush to about 2 metres tall. Its shiny, oval leaves are edged with obscure teeth and have a waxy appearance underneath. Other *Alseuosmia*s are shorter or sprawling. They have similar waxy undersides to their leaves, but leaf shape can vary from shrub to shrub. Their scented flowers are shorter than in karapapa, varying from crimson to pink, cream or greenish yellow. All *Alseuosmia*s have red oval berries, 5-10 mm long.

ENDEMIC GENUS

DOC

KAWHARAWHARA

## KAWHARAWHARA *(Astelia solandri)*

Kawharawhara is most commonly seen as a tufted plant with drooping, grassy leaves a metre or more long, growing high on forest trees. It also grows well on rocks or other well-drained, sunny sites out of the reach of browsing animals. Male and female plants are separate, and the flowers appear in summer on slender, branched, drooping, white woolly stems with pale leafy bracts. The flowers are clustered in cylindrical heads like bottlebrushes, and have six tepals, cream in females and pink to maroon in males. Females bear translucent yellow-green to brown berries in winter. Kawharawhara is the most widespread of several epiphytes with grass-like leaves in the genera *Astelia* and *Collospermum*, sometimes called 'perching lilies'. Kawharawhara and other tree-perching plants (epiphytes) are a feature of New Zealand's lowland forests, especially in wetter regions. They do not feed from the host tree, but are rooted in dead leaves and litter on branches or in tree forks. Other widespread epiphytes in New Zealand's forests include orchids, shrubs, ferns and some of the ancient relatives of ferns (*Lycopodium* and *Tmesipteris* species).

ENDEMIC SPECIES

Don Hadden

KOHUHU

## KOHUHU *(Pittosporum tenuifolium ssp. tenuifolium)*

Its naturally 'tidy' shape and wavy-edged leaves make kohuhu popular in gardens. It is also rewarding in late spring to seek out the deep wine flowers among the foliage, usually a single flower in the axil of each leaf. The flowers are scented, most strongly at night, and moths visit them. Pollination is also likely through birds, as bellbird and silvereye also visit kohuhu flowers for nectar. Opinions differ as to the degree to which kohuhu plants are functionally male or female. The five separate petals with out-turned tips make each flower about 8 mm across. Some flowers have pollen-bearing stamens as long as the style; others have sterile shorter stamens. Both kinds develop capsules of seeds, but some trees fruit more heavily than others. *Pittosporum tenuifolium* grows to about 8 metres, and is found from the coast to lower montane regions, but there are distinct regional forms. One, the common kohuhu of Westland, Stewart Island and the central North Island, is named subspecies *colensoi.* There is also regional variation in leaf size and shape, and leaf edge waviness in subspecies *tenuifolium*.

ENDEMIC SPECIES

Rod Morris

KORU

## KORU *(Colensoa physaloides)*

The flowers and fruits of koru make it one of the most remarkable native plants. The flowers are each 5 cm long, blue, and in sprays of 15-20 flowers. The berries are 25 mm long and blue or violet. Koru is a lush herb growing up to a metre tall. It is confined to northern islands and Northland, but even here it is often overlooked. Perhaps this is because it tends to live in secluded shady places in forest, especially beside streams or in damp gullies. The flowers are also often somewhat hidden by the lush foliage on one metre tall stems, woody at the base. Each flower is two-lipped, with the three lower petals usually paler blue than the upper two. The petals of each lip are partly joined, and all five petals are joined near the base. The stamen filaments are partly joined around the style, and the whole flower structure suggests that it is designed for bird pollination.

ENDEMIC GENUS

Rod Morris

KOWHAI

## KOWHAI *(Sophora microphylla)*

Its elegant form, soft green foliage and, above all, its lemon to gold flowers, make kowhai one of the best known and popular native trees in the bush and in cultivation. Kowhai is in the pea family, although its flowers may not suggest this at first sight. Nevertheless, the flower parts are arranged like those of kaka beak. Unlike kaka beak, kowhai petals are roughly equal in length, up to 45 mm long; the standard is curved outwards only a little, and the keel does not fully enclose the 10 separate stamens. The nectar attracts many birds, especially tui. Kowhai pods are constricted between each of the hard yellow or brown seeds. *Sophora microphylla* has many regional forms, and their differences remain under cultivation. As examples, Cook Strait kowhai can flower in May and are evergreen sprawling shrubs, eastern South Island kowhai are erect, deciduous trees to 10 metres, and weeping kowhai trees in the Rangitikei region flower in November. There is regional variation, too, in the size, number and hairiness of the leaflets. The larger-leaved *S. tetraptera* replaces *S. microphylla* in the northeastern North Island.

INDIGENOUS SPECIES

J & E Kelly

LANTERNBERRY, NOHI

## LANTERNBERRY *(Luzuriaga parviflora)*

Lanternberry grows in moist montane forests but descends to sea level in Fiordland and Stewart Island. It has very similar habitat needs and geographic range to the small forest iris *(Libertia pulchella)*, and the two often grow together. Lanternberry has creeping wiry stems that produce aerial, arching leafy shoots with zigzag stems to 250 mm long. Twisting of the leaf stalks (petioles) results in the leaves being inverted (the lower sides become uppermost) in two flattened rows. The dainty flowers are solitary and drooping, with six separate white tepals 10-18 mm long, forming an inverted cup. The six stamens have very short filaments attached around the base of the ovary. After seed set, the fruit becomes a white berry, almost as large as the flower. Not easily spotted much of the year, in summer lanternberry becomes quite conspicuous when the white flowers and fruit stand out against the moss and litter of the forest floor. The Southern Hemisphere plant family, Philesiaceae, has lanternberry as its one New Zealand member. Perhaps the best known relative of lanternberry is the Chilean flower, *Lapageria*.

ENDEMIC SPECIES

Don Hadden

MAHOE, WHITEYWOOD

## MAHOE *(Melicytus ramiflorus)*

Mahoe is one of the few forest trees whose flowers are more often found by scent than vision. Flowers of male trees are especially scented. The flowers are small, but large trees have so many that their perfume carries to the forest floor or well beyond the forest edge on a light breeze. This scent production may not be so surprising because mahoe is in the violet family. Flowering is in summer, in small clusters on young wood, just behind the leaves. Male flowers have five pale yellow petals and five anthers (the stamens lack filaments), and are about 5 mm across. Female flowers are slightly smaller, greenish-yellow, with a central ovary that eventually becomes a spherical purple berry (4-5 mm diameter) with several seeds. Mahoe is common and widespread in lowlands, ascending to montane forests in milder regions. Growing to 10 metres tall or more, its trunk tends to branch from near ground level and is knobbly and often hollow. Although mahoe grows in the understorey of tall forest, it is a primarily a canopy tree of regenerating forest or forest edges. It can be semi-deciduous.

ENDEMIC SPECIES

Don Hadden

MIKOIKOI

## MIKOIKOI *(Libertia grandiflora)*

*Libertia grandiflora* and its close relative, *L. ixioides*, grow in sunny or partly shaded gaps in lowland forest and shrublands. *L. grandiflora* flowers have six white tepals, the three inner ones twice the length of the outer and often completely covering them when viewed from above. Each flower has three stamens standing between the three style branches. As its formal name suggests, *L.. grandiflora* has large flowers (15-25 mm across), and these are on fine branched stems that are taller than the grassy leaves. Later, the small, three-part brown capsules split to release many small seeds. *L. ixioides* has smaller flowers on stems that are shorter than the leaves. This makes it less showy than *L. grandiflora*, although its capsules are larger and turn yellow or orange as they ripen. In different regions, *L. grandiflora* has either a single, tufted tussock form with 500-800 mm long leaves, or it grows in patches of plants connected by stolons. Future research may show that *L. grandiflora* is more than one species but, as understood at present, it is found as far south as Marlborough and Nelson.

ENDEMIC SPECIES

Rod Morris

NATIVE MISTLETOE

## NATIVE MISTLETOE *(Peraxilla colensoi)*

Flowering mistletoes, especially the red-flowered *Peraxilla* species, were once conspicuous, abundant and widespread in New Zealand forests. Browsing by introduced Australian possums has been shown as the major cause of mistletoe decline in many places. The decline has led to a great surge of interest in the conservation of native mistletoes in recent years, supported by research on their biology. Some of the most intriguing new findings have been on the structure and pollination of *Peraxilla* mistletoes. Mistletoe shrubs grow a metre or more in diameter and are semi-parasitic on other trees. *P. colensoi* has silver beech as its usual host, and is one of two native species with bright red flowers. Each flower bud has four narrow scarlet tepals, each with an anther attached inside, joined into a tube. Recent studies have shown that native birds, tui and bellbird, open ripe buds by 'tweaking' the bud tip, causing the sudden separation of the four tepals and giving access to the nectar. Flowers opened by birds are more likely to produce fruit than flowers which open unaided.

ENDEMIC SPECIES

Rod Morris

MOUNTAIN NEINEI, GRASS TREE

## MOUNTAIN NEINEI *(Dracophyllum traversii)*

With leaves like those of a cabbage tree, recurved and arranged in mop-top heads on candelabra branch ends, mountain neinei is one of the most distinctive trees in New Zealand. The flowers are each about 3 mm across with wine-coloured petals, arranged in dense conical heads about 300 mm long at the tips of leafy branches. The fruits are small dry capsules with many tiny seeds. In forest, the presence of mountain neinei is often first detected from the carpet of crisp fallen leaves on the ground, and the trunks with peeling grey or brown bark. Mountain neinei grows to 10 metres tall in montane, high rainfall forests in scattered North Island sites from near Kaitaia southwards to west of National Park, and on Little Barrier Island, and is common in parts of north-west Nelson, Westland and inland Canterbury. In the past, northern plants have been known as *D. pyramidale,* but this is generally now thought not to be distinct from *D. traversii.*

ENDEMIC SPECIES

Rod Morris

NIKAU

## NIKAU *(Rhopalostylis sapida)*

The only native palm of the New Zealand mainland, nikau is found in lowland forests as far south as Banks Peninsula in the east and Greymouth in the west. Nikau has a trunk 10 metres or more tall, distinctively ringed with the scars of fallen leaf bases. The feather-like fronds may reach 3 metres long. In the lowest leaf axils of mature trees the flower buds are crowded into branched heads, enclosed within two large pink bracts. When a leaf falls to expose the head of buds, the flower branchlets elongate and the bracts separate, often falling. Flowers grow mostly in groups of two males with one female, the males developing before females, in the same head. Each male flower is about 10 mm across, with three mauve petals and six stamens. Female flowers each have a mauve stigma on top of the ovary. The ovary becomes a single-seeded nut, 10 mm long, enclosed by a red fleshy coat when ripe. There is a different nikau species on the Kermadec Islands and the Chatham Islands nikau may be an unnamed species.

ENDEMIC SPECIES

COMMON GREENHOOD ORCHID, TUTUKIWI

# COMMON GREENHOOD ORCHID

*(Pterostylis banksii)*

Greenhoods are easily recognised as a group of ground orchids, but some individual species are quite difficult to distinguish. *Pterostylis banksii* grows in lowland forest. It is the largest and perhaps most widespread of the 27 or so native greenhoods. It grows to 50 cm tall, with grassy leaves and a single large flower at the stem tip. The flower is green with paler stripes and reddish fine points to sepals and petals. Two petals and one sepal are joined as a hood over the flower, the sepal having a long horizontal pointed tip. The other two sepals are joined and broad at their base, with long vertical points either side of the hood. As in many orchids, the third petal is a lip (labellum) designed as a landing stage for insects. In *P. banksii* and many other greenhoods, the lip has a trigger hinge, so that a visiting insect is thrown into the bottom of the flower and, as it climbs out, it brushes any pollen it was carrying on the stigma, accidentally picks up new pollen, and emerges to visit another orchid.

ENDEMIC SPECIES

DOC

EASTER ORCHID, RAUPEKA

## EASTER ORCHID *(Earina autumnalis)*

Flowering Easter orchids are often found in the bush by their sweet rich perfume. They can flower between February and July, depending on climate and altitude, but usually around Easter. The pure white flowers with a yellow-orange patch on the lip (labellum) are about 13 mm across, made more conspicuous by growing close together in spikes of up to 40 flowers. Close study of a flower shows its shape to be very similar to that of the lady's slipper orchid. The Easter orchid is a common epiphyte through much of New Zealand, often growing with its relative *E. mucronata* and the lady's slipper orchid. Such epiphytes do not feed from the host trees but merely perch on them as places with good light and drainage. Easter orchids have tough grassy leaves that are twisted so as to appear in two rows. The unbranched wiry stems, to 1 metre, usually hang from tree branches, but the flowering tips curve upwards. Easter orchids can grow as large patches on well drained ground in shrublands and open beech forest, and they also grow on rocks and cliffs.

ENDEMIC SPECIES

Rod Morris

LADY'S SLIPPER ORCHID, WINIKA

# LADY'S SLIPPER ORCHID

*(Dendrobium cunninghamii)*

*Dendrobium* is an orchid genus of more than 1000 species world-wide, of which *D. cunninghamii* is the sole New Zealand species. Most of the 110 or so New Zealand orchids are purely terrestrial plants, but, of seven predominantly epiphytic species, lady's slipper has by far the largest flowers, up to 30 mm across. Flowers have three sepals and two petals, which are white and similar in size. In different parts of the orchid's range, the lip (labellum) may be pure white or, more often, has mauve and yellow patches, and there is some variation in flower shape and size. Many native orchids are self-pollinated, but experimental work shows the lady's slipper orchid has to be cross-pollinated, probably by insects, to produce seeds. Lady's slipper orchid is common in wetter parts of the country, usually growing on the branches of tall forest trees as a branched bushy shrub a metre or more across. The branched yellowish stems resemble fine bamboo. Because trees provide only physical support and good drainage for its roots, lady's slipper orchids sometimes grow on cliffs or rock outcrops.

ENDEMIC SPECIES

Rod Morris

POROPORO

## POROPORO *(Solanum aviculare, S. laciniatum)*

Many people are surprised to learn that poroporo is a native plant, for its purplish flowers, bold against dark green foliage, are quite unlike flowers of any other native shrub. Poroporo grows as a soft shrub to 3 metres, and is common in weedy places such as waste areas, rough grassland and areas of young regeneration. However, it is also a plant of coastal and lowland forests, especially around the edges and in canopy gaps. There are two species of poroporo, with very similar foliage, form and habitats, but easily distinguished by the size and colour of their flowers. *Solanum laciniatum* has bright blue-purple petals, joined in a flattened bell shape 40-50 mm across, with wavy edges. The pale purple or whitish petals of *S. aviculare* are joined in a deeper cup 30-35 mm across, with five pointed petal tips. Poroporo has five stamens in each flower, with large, bright yellow anthers that are pressed together around the style until their pollen is shed through pores at their tips. The ripe berries are egg-shaped, pale yellow and 25-30 mm long in *S. laciniatum*, and more nearly spherical, orange and 20-25 mm long in *S. aviculare*.

INDIGENOUS SPECIES

Rod Morris

PURIRI

## PURIRI *(Vitex lucens)*

Puriri is one of the most splendid native trees for its solid, often gnarled, form and its dark, glossy broad crown. It is also among the most important forest trees for birds, because it has nectar-filled flowers and fleshy fruits much of the year, and most abundantly in winter. Puriri has clusters of 5-15 flowers drooping near the tips of leafy branchlets. Despite often being obscured by the leaves, the flowers are quite striking, being 25-35 mm long with a two-lipped tube of petals, pale cream but usually flushed with pink or deeper pink to pink-red. The style and the four stamen filaments are curved and lie inside the curve of the upper petal lip. This puts the anthers in a group beyond the petals, well-placed to dust their pollen on the heads of visiting birds. Puriri is a lowland forest tree, or mostly coastal and semi-coastal near its southern limits in North Taranaki and near Gisborne. It grows to 20 metres, typically in fertile, free-draining soils.

ENDEMIC SPECIES

J & E Kelly

PUTAPUTAWETA, MARBLELEAF

## PUTAPUTAWETA *(Carpodetus serratus)*

In flower, putaputaweta is one of the most eye-catching smaller trees of the bush. The starry white flowers are displayed in bunches on the upper side of the horizontal branches and branchlets that give a layered look to the tree. Viewed from above, a flowering tree is especially striking. Each five-petalled flower has male and female parts and is up to 10 mm across. There is some evidence that smaller flowers are functionally female only and occur on the most heavily fruiting trees. Flowers are followed by hard capsules, black when ripe. Putaputaweta grows up to 10 metres tall through much of New Zealand, from near the coast to montane regions. It tends to be in wet forest such as valley floors and swamps. Putaputaweta leaves resemble those of red or hard beech, but they have a mottled pattern of pale and darker green which earned the tree the English name 'marbleleaf'. In the North Island, the trunk is frequently riddled with holes eaten by puriri moth larvae. Wetas (native crickets) occupy empty holes, and gave rise to the name 'putaputaweta' (full of weta holes).

ENDEMIC SPECIES

Don Hadden

RANGIORA

## RANGIORA *(Brachyglottis repanda)*

At first sight, rangiora, a shrub up to 6 metres tall, may not look like a daisy, but a close study of its flowers shows they are grouped in the characteristic heads (capitula) of the daisy family. Each rangiora capitulum is about 5 mm across and contains some 10-12 tiny daisy flowers (florets). As in the case of many daisies, the seeds that follow the flowers are attached to tufts of fluff like thistle-down, for wind dispersal. Although the capitula are small individually, they grow in large bunches (panicles) of hundreds of heads. These make flowering shrubs of rangiora quite conspicuous at a distance, for they commonly grow on forest edges, young erosion scars, roadside banks, and in shrublands. On hot calm days, the musky scent of the flowers is very distinctive. When not in flower, a rangiora shrub is easily recognised by its large oval leaves, often broader than a person's hand. These leaves have white-felted undersides, dull green upper surfaces, and inconspicuous blunt lobes around the margins. The younger stems are also covered in white matted hairs.

ENDEMIC SPECIES

Key-Light

CLIMBING RATA, AKA

## CLIMBING RATA *(Metrosideros perforata)*

*Metrosideros perforata* is a vine of coastal and lowland forests as far south as Banks Peninsula in the east and South Westland. Its foliage and flowers are most easily seen on forest margins, but it can climb to 15 metres or more in the forest interior. On the coast or exposed ridges *M. perforata* can grow as a tangled bush no more than 1 metre tall. Like other species of *Metrosideros*, *M. perforata* has flowers in which the showy parts are the stamens. The 30 or more stamens in each flower are white or pink flushed, 8-10 mm long, and arranged around the rim of a hollow disc that contains nectar. The five round, white or pale pink petals are barely 1 mm long. Clusters of flowers are crowded at the stem tips, often smothering the dark green foliage. It has been stated that more insects visit the flowers of this rata than any other native plant. The dense flowering of *M. perforata* is more striking than the diffuse flowering of the other widespread white rata vines, *M. diffusa* and *M. colensoi*, which often grow with it.

ENDEMIC SPECIES

Rod Morris

SCARLET RATA VINE, AKATAWHIWHI

## SCARLET RATA VINE *(Metrosideros fulgens)*

The scarlet or orange flowers of akatawhiwhi are all the more striking because they appear in late summer and autumn when there are few other native flowers in the bush. It is similar to other species of *Metrosideros* in having stamens as showy parts of its flowers. Each flower has five blunt sepals about 3 mm long, and five rounded orange-red petals that are slightly longer. These cover 50 or more incurled stamens in the bud stage. As the flower opens, the stamens straighten and elongate to 20-25 mm, around the rim of a shallow disc with nectar. Dense clusters of up to 12 flowers grow at the ends of leafy branches, their visual impact being greatest when the akatawhiwhi vine grows masses of leafy branches over shrubs or low trees on the forest edge. In the heart of the bush, akatawhiwhi grows as a thick (100 mm diameter) rough-barked vine, to 10 metres tall. Akatawhiwhi is found on the coast and in lowlands of the North Island and the north and west of the South Island.

ENDEMIC SPECIES

Rod Morris

SOUTHERN RATA

## SOUTHERN RATA *(Metrosideros umbellata)*

Southern rata is a round-headed tree, to 15 metres or more, with red flowers. To see the massed flowering of southern rata forest can be one of New Zealand's most spectacular forest experiences. This is probably best seen against Westland's dramatic landscapes. Because southern rata grows from the coast to subalpine zones, flowering can start in December and, with increasing altitude and distance south, in March or even later. As in other species of *Metrosideros,* the conspicuous parts of the flowers are the stamens. These are scarlet, or sometimes crimson or orange. Flowers grow in clusters at the ends of leafy shoots. Each flower has five small sepals, five red petals about 5 mm long, and many 20 mm stamens around the raised rim of a nectar-containing cup. Southern rata can be the dominant tree in the western and southern South Island, Stewart and Auckland Islands but it is also in some elevated wet parts of the North Island, from near Kaitaia southwards, and isolated, sometimes quite dry, places in the eastern South Island. Its abundance has been reduced by the browsing, introduced Australian possum.

ENDEMIC SPECIES

Brian Chudleigh

REWAREWA, NZ HONEYSUCKLE

## REWAREWA *(Knightia excelsa)*

Rewarewa is one of the two New Zealand members of the ancient *Protea* family. Viewed close-up, its open flowers resemble those of its Australian relatives such as *Grevillea* and *Hakea*, and are borne in loose cylindrical heads like those of the Australian genus *Banksia*. Each rewarewa flower bud has four tepals joined into a rich red-brown tube. As the flower opens, the tepal tips split apart, revealing the anthers attached to their inside surfaces. The anthers deposit pollen on the immature stigma at this stage. The tepals then split along their whole length and coil rapidly to the flower base, taking the anthers with them. The stiff styles remain as a kind of bottlebrush, and nectar-feeding birds take pollen off the stigmas as they push between the styles. After the stigmas ripen, bird visitors pollinate the flowers with pollen carried from other flowers. Rewarewa fruits are dry capsules containing four winged seeds. Rewarewa is found throughout lowland and lower montane forests of the North Island and Marlborough Sounds, often as a cylindrical tree, to 30 metres tall, in regenerating forest or shrublands.

ENDEMIC SPECIES

DOC/C.R. Veitch

MOUNTAIN RIBBONWOOD

## MOUNTAIN RIBBONWOOD *(Hoheria lyallii)*

Mountain ribbonwood in flower looks rather like a leafy, flowering apple tree. Each flower is about 40 mm across, with five white petals that overlap, the flower having the shape of a shallow saucer. In the centre are five bunches of stamens and 10-15 styles, each attached to a carpel of the ovary. The carpels separate when ripe into dry one-seeded fruits. Mountain ribbonwood is an erect, slender, deciduous tree up to 10 metres tall. It grows in rocky gullies, forest edges, river banks and stony erosion scars, sometimes in lowlands but mostly in montane zones. It sometimes forms groves on stony or tussock-covered mountain slopes where no other trees occur. As originally defined, *H. lyallii* was a tree of the eastern South Island, but it is now widely believed that it and *H. glabrata*, the mountain ribbonwood of the western South Island and Mt Taranaki/Egmont are one species, known collectively by the earlier name, *H. lyallii*. The two species were distinguished mostly by the quantity of stellate (star-shaped) hairs on the leaves and twigs.

ENDEMIC SPECIES

Don Hadden

TARATA, LEMONWOOD

## TARATA *(Pittosporum eugenioides)*

Flowering tarata is sometimes found by its sweet honey perfume before the pale yellow flowers are seen. Some trees are effectively female, for the flowers have a large ovary that develops a fruit, and five non-functional stamens with the anthers on short filaments. Other trees have flowers with five functional stamens on long filaments and only a small ovary which may sometimes develop fruit. The fruit is a dry capsule about 5 mm long that splits into two or three parts when ripe. Tarata can be a broad-crowned canopy tree of 12 metres tall with a trunk over 50 cm across, but is often seen as a smaller tree or shrub on forest margins, river banks or in regenerating forest. It is a popular garden shrub for its naturally clipped shape and yellow-green oval leaves with wavy edges. Its natural range is throughout New Zealand in lowlands or lower montane forests, usually in places with high fertility soils, such as river flats. Crushed leaves have a lemon scent, hence its common name 'lemonwood'.

ENDEMIC SPECIES

Brian Chudleigh

TAWARI

# TAWARI *(Ixerba brexioides)*

Tawari is one of the most splendid native trees in flower. It grows to 10 metres tall with spreading branches that give a broad canopy of dark green leaves, against which the panicles of flowers can be seen at considerable distances. Each flower is about 30 mm across, and has five white petals with narrow bases, the gaps between being filled by green-white sepals about half the petal length. Five stamens stand out beyond the petals and around the central ovary is a flattened disc that exudes copious nectar. Flowering is between November and January, depending upon location, for tawari grows from near the coast to about 900 metres in altitude. After flowering, five-part capsules split open, each part releasing two glossy black seeds. Tawari tends to grow in low-fertility soils, and can be the dominant tree on forested ridges or the sides of gorges. It is found in the North Island only, from near Puketi Forest in Northland to the central King Country. The genus *Ixerba* has one species.

ENDEMIC GENUS

DOC

WHAU

## WHAU *(Entelea arborescens)*

Whau is one of the most distinctive native trees, not least for its bunches of large (25 mm across) flowers that can obscure much of the foliage of the tree crown in late spring. Each flower has four or five narrow white sepals and pleated white petals with slightly wavy edges. No other native tree has spine-covered capsules like those of whau. Whau is frost-tender, but in coastal forest it can grow faster than perhaps any other native tree, to take advantage of new gaps in the forest canopy. The shallow-rooted trees, to 6 metres tall, fall over readily, but these often coppice to form thickets of erect trunks. Whau leaves are heart-shaped, soft and drooping, with toothed edges, usually about 150-200 mm across but can be much larger in the shade. Because whau's very light-weight wood was used by the Maori for items such as fishing net floats, the natural range of whau may have been extended by early plantings. It is known as far south as North Taranaki, near Wellington, and in the north of the South Island.

ENDEMIC GENUS

Don Hadden

WINEBERRY, MAKOMAKO

## WINEBERRY (*Aristotelia serrata*)

Wineberry is an erect shrub or tree, to 10 metres high, widespread on forest edges and in canopy gaps in the lowlands and mountains, especially where rainfall is high. Although its flowers are only about 5 mm across, wineberry has a massed flowering in spring that often brings it to people's notice. Its flowering is also obvious because it establishes and flowers quickly in disturbed places such as road verges, erosion scars and canopy gaps in the forest. Wineberry and hinau are in the same plant family but the former has male and female flowers on separate trees. Even though they differ greatly in size, the flowers of wineberry and hinau are very similar in shape, especially in the lobed edges of the petals. Female flowers are followed in summer by red or black berries, 5 mm across. *Aristotelia fruticosa*, or mountain wineberry, is a twiggy shrub of lowland to montane forest and shrublands. At first sight, it looks quite unlike wineberry but its flowers are very similar and hybrids between the two species are sometimes found in the wild.

ENDEMIC SPECIES

# SCIENTIFIC CLASSIFICATION

All animals and plants are grouped and classified according to their structure, shape, physiology, distribution and mode of life. In this way relationships between groups of living things can be established.

The basic single natural unit in this classification system is the *species*, a group whose members in the wild can interbreed and produce fertile offspring. Closely related species form a *genus*. Related genera form *families*, and families *orders*. Orders are grouped into *classes*, classes into *phyla*, and phyla are grouped into either the animal or plant kingdoms. In giving a species its scientific classification, a generic name, representing the name of a group of related species (or genus), is followed by the specific name of the particular species. The names are in Latin, and the two together are known as the species' scientific name. In the case of a subspecies, a third or trinomial name follows the specific name.

A subspecies occurs where part of the population within a species shows distinct differences in appearance — in size, shape or colour. Subspecies, or races, usually also occupy a distinct geographical area, often as isolated populations on islands where they have evolved in isolation from other groups of the same species. In New Zealand with its numerous offshore and outlying islands and steep, confining, mountainous terrain, a considerable number of subspecies have evolved.

The following list shows the flower species presented in this book grouped by their scientific classification.

**Subdivision Angiospermae** (flowering plants)

1. **DICOTYLEDONS:** Flowering plants whose seedlings have two (rarely more) seed leaves (cotyledons); leaves mostly broad and net-veined; flowers with parts in fours or fives.

| ***ORDER*** | ***FAMILY*** *(with other examples)* | ***GENUS AND SPECIES IN THIS BOOK*** |
| --- | --- | --- |
| **Ranunculales** | Ranunculaceae *(buttercups, Anemone, Aquilegia)* | *Clematis paniculata* |
| **Malvales** | Elaeocarpaceae (*pokaka*) | *Aristotelia serrata*<br>*Elaeocarpus dentatus* |
| | Tiliaceae (*limes*) | *Entelea arborescens* |
| | Malvaceae *(mallows, cotton, Hibiscus)* | *Hoheria lyallii* |
| **Violales** | Violaceae (*violet*) | *Melicytus ramiflorus* |
| **Fagales** | Fagaceae (*beeches, oaks, sweet chestnut*) | *Nothofagus solandri* var. *cliffortioides* |
| **Cunoniales** | Cunoniaceae (*Cunonia, makamaka*) | *Weinmannia racemosa* |
| **Rosales** | Rosaceae (*rose, apple, peach, blackberry*) | *Rubus cissoides* |
| **Protealaes** | Proteaceae (*Protea, Banksia*) | *Knightia excelsa* |
| **Myrtales** | Myrtaceae (*myrtles, pohutukawa, manuka, Eucalyptus, bottlebrushes*) | *Kunzea ericoides*<br>*Metrosideros fulgens*<br>*Metrosideros perforata*<br>*Metrosideros umbellata* |
| | Onagraceae (*willow herbs, evening primroses*) | *Fuchsia excorticata* |
| **Fabales** | Fabaceae (*legumes, wattles*) | *Clianthus puniceus*<br>*Sophora microphylla* |

| *ORDER* | *FAMILY* (with other examples) | *GENUS AND SPECIES IN THIS BOOK* |
|---|---|---|
| **Santalales** | Loranthaceae (*mistletoes*) | *Peraxilla colensoi* |
| **Pittosporales** | Pittosporaceae (*karo*) | *Pittosporum eugenioides*<br>*Pittosporum tenuifolium* |
| **Campanulales** | Lobeliaceae (*Lobelia*) | *Colensoa physaloides* |
| **Asterales** | Asteraceae (*daisies*) | *Brachyglottis repanda*<br>*Olearia arborescens* |
| **Solanales** | Solanaceae (*nightshades, potato, Datura, tamarillo*) | *Solanum aviculare*<br>*Solanum laciniatum* |
| **Ericales** | Epacridaceae (*Epacris, Cyathodes, Leucopogon*) | *Dracophyllum traversii* |
| **Cornales** | Escalloniaceae (*Escallonia*) | *Carpodetus serratus*<br>*Ixerba brexioides* |
| | Alseuosmiaceae (*small family of NZ, New Caledonia, and possibly Australia & PNG*) | *Alseuosmia macrophylla* |
| **Scrophulariales** | Gesneriaceae (*Gloxinia*) | *Rhabdothamnus solandri* |
| | Scrophulariaceae (*foxglove, Veronica, Hebe*) | *Jovellana sinclairii* |
| **Lamiales** | Verbenaceae (*Verbena, teak, Lantana, Teucridium*) | *Vitex lucens* |

2. **MONOCOTYLEDONS:** Flowering plants whose seedlings have one seed leaf (cotyledon); leaves mostly linear (grass-like) with parallel veins; flowers with parts in threes.

| *ORDER* | *FAMILY* (with other examples) | *GENUS AND SPECIES IN THIS BOOK* |
|---|---|---|
| **Alstromeriales** | Philesiaceae (*Lapageria*) | *Luzuriaga parviflora* |
| **Liliales** | Asphodelaceae (*asphodel*) | *Astelia solandri* |
| **Iridales** | Iridaceae (*irises, Gladiolus, Warsonia*) | *Libertia grandiflora*<br>*Libertia pulchella* |
| **Palmales** | Palmaceae (*palms*) | *Rhopalostylis sapida* |
| **Orchidales** | Orchidaceae (*orchids*) | *Dendrobium cunninghamii*<br>*Earina autumnalis*<br>*Pterostylis banksii* |

INDEX